History & Future of Cubesats

by Patrick H. Stakem
(c) 2021

Number 5 in the Cubesat Series

Table of Contents

Introduction

This books describes the history and future of Cubesats, a paradigm shift in space exploration. As of this writing, some 1,350 units have been launched. The STMSat-1 cubesat was launched from the International Space Station in December of 2015. It had been built by students of the Saint Thomas More grade school in Arlington, Va. From University-defined, low cost satellite, the Cubesat concept is widely having an impact on the way we do exploring in space. They have been to the Moon and Mars. They are small, but mighty, and produce cost savings not only due to their size, but also their modularity. They are, essentially off-the-shelf platforms that can host a wide variety of instruments, and just wait until you see the collection of cooperating units doing science.

To date, more than 1,350 Cubesats have been launched. Space-X holds the record for most satellites launched in one flight, at 143. Of these, 120 were Cubesats.

Cubesats can do ride-sharing, going along on a mission that has some space and weight margins.

Author

The author has a BSEE in Electrical Engineering from Carnegie-Mellon University, and Masters Degrees in Applied Physics and Computer Science from the Johns Hopkins University. During a career as a NASA support contractor from 1971 to 2013, he worked at all of the NASA Centers. He served as a mentor for the NASA/GSFC Summer Robotics Engineering Boot Camp at GSFC for 2 years. He taught Embedded Systems for the Johns Hopkins University, Engineering for Professionals Program, and has done several summer Cubesat Programs at the undergraduate and graduate level.

What is a Cubesat?

A Cubesat is a small, affordable satellite that can be developed and launched by colleges, high schools, and even individuals. The specifications were developed by Academia in 1999. The basic structure is a 10 centimeter cube, (volume of 1 liter) weighing less than 1.33 kilograms. This allows multiples of these standardized packages to be launched as secondary payloads on other missions. A Cubesat dispenser has been developed, the Poly-PicoSat Orbital Deployer, P-POD, that holds multiple Cubesats and dispenses them on orbit. They can also be launched from the Space Station, via a custom airlock. ESA, the United States, and Russia provide launch services. The Cubesat origin lies with Prof. Twiggs of Stanford University and was proposed as a vehicle to support hands-on university-level space education and opportunities for low-cost space access. This was at a presentation at the University Space Systems Symposium in Hawaii in November of 1999.

Cubesats began as teaching tools, and remain in that role, although their vast numbers in orbit showed they they have become mainstream.

The Cubesat Design Specification

The Cubesat Design Specification, developed by California Polytechnic State University, defines the physical and interface specifications for Cubesats, and gives testing requirements for vibration, thermal-vacuum tests, and shock, as well as safety. Since a Cubesat flys with other Cubesats in a deployment device, and with a primary payload, safety is a concern. Cubesats are expected to have an on-orbit lifetime of less than 30 years.

The first grade school to put a Cubesat (STMSat-1) into orbit was the St. Thomas More Cathedral School in Arlington, Virginia, in 2012.

Here is a synopsis of CubeSat requirements:

Mass

- Each satellite may not exceed 1 kg of mass.

- The CubeSat center of mass must be within 2 cm of the geometric center.

Structure

All edges contacting rails must be rounded. Cubesats must have at least 75% (85.125 mm of a possible 113.5 mm) of flat rail contact with the deployer- To prevent cold-welding, raw metal is not allowed as the contact surface of the bottom standoff. Derlin inserts, or a hard anodize are examples of acceptable contact surfaces.

The outer surfaces of the CubeSats are required to be hard anodized in order to prevent wear between the sliding rails and the CubeSats.

Separation springs (SSMD-51P recommended) must be included at designated contact points. A custom separation system may be used upon approval by CalPoly/Stanford launch personnel. One deployment switch is required (two are recommended) for each CubeSat.

Material

The use of Aluminum 7075 or 6061-T6 is suggested for the main structure. If other materials are used, the thermal expansion coefficient must be similar to that of Aluminum 7075-T3 (the POD material) and approved by CalPoly/Stanford personnel.

Deployables

A time delay, on the order of several minutes, must be present between release from the P-POD and any satellite hardware deployment, to allow for satellite separation.

P-POD rails and walls cannot be used to constrain deployable hardware

Communication

There must be a time delay, on the order of several minutes to an hour, before primary transmitters are activated. Low power beacon transmitters may be activated after deployment.

Operators must provide proof of the appropriate license for radio frequency use.

Power

CubeSats with rechargeable batteries must have the capability to receive a transmitter shutdown command, compliant with FCC regulations.

Satellites that require testing and battery charging must provide an external hardware interface to access the power/data port

A 'remove before flight' pin is required to deactivate the CubeSats during integration outside the P-POD. The pin will be removed once the CubeSats are placed inside the P-POD.

General

Absolutely no pyrotechnics are allowed inside the CubeSat.

Cubesats are defined by Units, where a unit is 10 x 10 x 10 Centimeters. Multiples of units are also allowed.

There are emerging standards for larger Cubesats, such as 6U (12 kg, 12 x 24 x 36 cm), 12U (24 kg, 23 x 24 x 36 cm) and 27U (54 kg, 34 x 35 x 36 cm). These allow the canister to constrain Cubesat deployables such as antennae and solar array. In the original Cubesat specification, this task had to be handled by the Cubesat itself. Even as they get bigger, the standard architecture and modularity of the Cubesat remains a game-changing advantage.

If you're going to launch a Cubesat into Earth orbit, you have to carefully follow the Cubesat Design Specification. Where's the fun in that? We will introduce a concept here which will allow a Cubesat to be put on various mobility platforms, and visit other planets.

In fact, if you forget about space for the moment, you can bolt a Cubesat onto a variety of mobility platforms, that can explore Earth from the sky, from under the Antarctic ice, and in Lava Tubes.

Lava tubes on the Moon and Mars are of interest, as they might provide the basis of a protected, habitable volume for astronauts.

We will show how "Cubesats" can operate together cooperatively to tackle big, complex problems. This is the "Strength in Numbers" Scenario. We have shown that, within the envelope of one large, expensive exploration spacecraft to Jupiter, you can send 1,000 Cubesats. For something like the Asteroid Belt, where you have 10's of thousands of potential targets, there is no practical solution to explorationexcept a swarm of small satellites.

Space environment

Now, we'll discuss the environment in which the Cubesat will operate, assuming it survives launch. It is not going to sit on your bench or desk. It's going to space.

The space environment is hostile and non-forgiving. There is very little gravity, so no convection cooling is possible, leading to potential thermal problems. It is in a high radiation environment, being above the shielding provided by the atmosphere. The

Cubesat system is power constrained, and, it is hard to debug and repair after launch, as with almost all missions.

There are differing environments by Mission type. For Near-Earth orbiters, there are the radiation problems of the Van Allen belts and the South Atlantic Anomaly, the thermal and vacuum environment, and the issue of atmospheric drag. This drag causes orbital decay, where the spacecraft slowly descends. There is also a drag factor from the residual atmosphere and the solar wind, and the spacecraft's orbit can be affected in other ways. Almost all Cubesat missions are currently near Earth, although NASA has developed specialized Cubesats for planetary exploration. Two of these are past Mars, and are heading out further.

PiSat

The PiSat is an open source NASA/GSFC design for a "Distributed Mission Test Platform." It represents an ideal platform for prototyping Cubesat Flight software, as well as educational outreach. The PiSat defines the flight computer (ARM), a sensor suite, the enclosure and battery, and the Flight Software. It was developed by NASA/GSFC Code 582, with IRAD funding. It was developed with help from undergraduate student interns.

The Flight Computer is the Raspberry Pi (2 B+), based on the ARM architecture, and running the linux operating system or RTOS, with Code 582's Core Flight System software suite. The supported sensors include a GPS module, magnetometer, compass, accelerometer, a high definition camera, A/D converters, and a real-time clock. Data storage is provided by an SD flash memory card. It uses the Xbee peer-to-peer wireless communication. The price of the hardware components comes in around $350, including the printed enclosure.

Cubesats accept the PC-104 board standard (90 mm x 96 mm), and the boards are stackable. There is no requirement to use this size board, or the standard, but there can be advantages, such as availability of interfaces. The unit is powered via USB during test and development, and by a standard lithium battery for flight.

The CFS software is reusable mission software that has already flown on many NASA missions, including the Lunar Reconnaissance Orbiter, and covers common onboard tasks. A collection of applications under the CFS includes uplink and downlink of data, attitude calculation, and support of the camera. There are a set of scripts for startup and shutdown of the system. For test or operations, there are several software choices, including the COSMOS system from Ball Aerospace.

An intern (and student of the author) integrated an ocean spectrometer into the PiSat architecture at GSFC.

Launch Environment

Even without a launch vehicle failure, the launch environment represents the worst case in the path to orbit. There is significant shock and vibration, and a large acoustic environment. This only lasts for a few minutes, but is severe. The spacecraft must be tested beyond the expected launch environment, data for which is available from the launch vehicle provider. The Thermal/Vacuum and Vibration/Acoustics tests are usually referred to as "Shake and Bake."

In addition, we need to examine the structure of the Cubesat to ensure that it is properly vented, so the residual internal atmosphere can vent during ascent.

Zero G issues

Zero gravity, actually, free-fall, brings with it problems. There is no convection cooling, as that relies on the different densities of warm and cool air. Any little pieces of conductive material will float around and short out critical circuitry at the worst possible time. And then, there are the strange issues.

Vacuum

The Cubesat operates in vacuum. Not a perfect vacuum, but fairly close. This implies a few things. Lubricants disappear. All the materials out-gas to some extent. All this material can find its way to condense on optical surfaces.

Thermal environment

In space, things are either too hot or too cold. Cooling is by conduction to an outside surface, and then radiation to cold space. This requires heat-generating electronics to have a conductive path to a radiator. That makes board design and chip packaging complex and expensive. You get about 1 watt per square meter of sunlight in low Earth orbit. This will heat up the spacecraft, or you can convert it to electrical power with solar arrays.

Parts can be damaged by excessive heat, both ambient and self-generated. In a condition known as *thermal runaway*, an uncontrolled positive feedback situation is created, where overheating causes the part to further overheat, and fail faster.

There can be a large thermal gradient of hundreds of degrees across the satellite, where one side faces the sun, and the other side faces cold space.

History

A Cubesat is a small, affordable satellite that can be developed and launched by college, high schools, and even individuals.

Cubesats began as teaching tools, and remain in that role, although their vast numbers in orbit showed they they have become mainstream.

In what has been called the Revolution of smallsats, Cubesats lead the way. They represent paradigm shifts in developing space missions, opening the field from National efforts and large Aerospace contractors, to individuals and schools.

Initially, Cubesats were launched as "ride-sharing" units on a regular payload mission. It is not widely known, but the center of mass of the payload is frequently adjusted with bags of sand. Now, it is adjusted by Cubesats. A revolutionary mission launched over 100 Cubesats to orbit, with no primary payload. Cubesats have been to the Moon, and Mars. They are a different way of thinking about exploration of Space, or exploration in general.

Disposal

By requirement, a Cubesat is required to reenter the Earth's atmosphere within 25 years. This is to avoid contributing to the orbital debris Problem. Obviously if the Cubesat mission takes it away from the Earth, either towards the Sun or towards the outer planets, this restriction does not apply. Also, Cubesats delivered to the International Space Station can remain attached to the outside of the station for long periods of time, and be retrieved and returned by a resupply flight.

Propulsion

Generally, Cubesats are not allowed to have a propulsion system. They are specifically constrained to not exhaust any gas, even nitrogen. However, NASA can use propulsion on their own small satellites. Another way for Cubesats to wander the solar system is via solar sails, which use the solar wind, emanating from the Sun, much as a sailboat uses the terrestrial wind. This has been demonstrated with Cubesats.

Not only just one.

There is strength in numbers. If you think a Cubesat-based payload can get a it done, you look at what a thousand of them can do. Here we discuss aggregations of Cubesats. They may interact with each other. Some of the architectures include Trains, Constellations, Clusters, and Swarms. An interesting point is, not all the members of the groups need to be launched or deployed at the same time.

Beyond deploying Cubesats as ride-along payloads, we can postulate a very large Cubesat carrier with hundreds or thousands of Units, being delivered to a target of interest. The Cubesats here are the primary payload. The Constellations can be "grown" and expand as the mission progresses, and failed units can be discarded and replaced.

Trains

Trains of satellites refer to multiple units that are spaced along the same orbital track. This allows for simultaneous imaging areas, as well as continuous observation of selected areas. It is a co-ordinated group of observation satellites. NASA uses this approach successfully for Earth Science and Weather satellites in polar, sun-synchronous orbits. A train of 6 weather satellites passes over the same spot at the same time every day. The satellites are all different, but provide useful information on atmospheric and ground conditions. Another term for a simple group of spacecraft in the same orbit is *string-of-pearls*.

Complexity in a system generally derives from two parameters, the number of units, and the number of interactions. So a swarm of cubesats is complex, compared to a single spacecraft. This is somewhat balanced by the relative individual simplicity of the units, and their flexibility and redundancy.

Constellations

Constellations are groups of satellites operating together to observe a single target A constellation allows you to do simultaneous observations of one target from multiple locations, or multiple targets simultaneously. The elements of a constellation can be homogeneous or diverse.

NASA says, " A Constellation is a space mission that, beginning with its inception, is composed of two or more spacecraft that are placed into specific orbit(s) for the purpose of serving a common objective (e.g., Iridium)." (The Iridium constellation is for

13

communications).

An actual Constellation of 50 2U and 3U Cubesats was deployed in 2015. Some were released from the ISS, and some from a rocket launch. They collected and telemetered data on the lower thermospere. This was not a Constellation, per se, but 50 units acting on their own, reporting back to their home institutions. Universities around the world participated, and built units from the QB50 specification.

The Distributed Spacecraft Mission was defined at JPL to allow "formation flying" of multiple spacecraft. Another point of view is the "fractionalized spacecraft", where the spacecraft functionality is distributed across multiple units. Critical to this is a intra-communications mechanism. The Constellation may use a mesh or lattice architecture. The members of this organization can be launched together, or separately.

A Cubesat constellation was formed in the Flock-1 project in 2014. It consisted of 28 units. It was launched from the Mid-Atlantic Regional Spaceport to the ISS. They were deployed to orbit in February 2014. The mission was Earth observation with a camera capable of resolving objects 3 meters on the surface.

Swarms

This section describes a different approach: collections of smaller co-operating systems that can combine their efforts and work as ad-hoc teams on problems of interest. Cubesats can be organized in Swarms.

This is based on the collective or parallel behavior of homogeneous systems. This covers collective behavior, modeled on biological systems. Examples in nature include migrating birds, schooling fish, and herding sheep. A collective behavior emerges from interactions between members of the swarm, and the environment. The resources of the swarm are organized dynamically.

A driver in the space environment is the exploration of the asteroids. Although there are fewer than 10 planets, and less than 200 moons, there are millions of asteroids, mostly in the inner solar system. The main asteroid belt is between Mars and Jupiter. Each may be unique, and some will provide needed raw materials for Earth's use. There are three main classifications: carbon-rich, stony, and metallic.

The physical composition of asteroids is varied and poorly understood. Ceres appears to be composed of a rocky core covered by an icy mantle, whereas Vesta may have a nickel-iron core. Hygiea appears to have a uniformly primitive composition of carbonaceous chondrite. Many of the smaller asteroids are piles of rubble held together loosely by gravity. Some have moons themselves, or are co-orbiting binary asteroids. The bottom line is, asteroids are numerous and diverse.

It has been suggested that asteroids might be used as a source of materials that are rare or exhausted on earth (asteroid mining) or materials for constructing space habitats or as refueling stations for missions. Materials that are heavy and expensive to launch from Earth may someday be mined from asteroids and used for space manufacturing. Valuable materials such as platinum may be returned to Earth for a profit.

Exploring the asteroids requires a diverse and agile system. Thus, a swarm of robotic spacecraft with different capabilities might be used, combining into Teams of Convenience to address situations and issues discovered in situ.

Biological swarms, such as ants, achieve success by division of labor throughout the swarm, collaboration, and sheer numbers. They have redundancy, as any individual can do any task assigned to the swarm. The individual units are highly autonomous, but are dependent on other members for their needs. They achieve success with a simple neural architecture and primitive communications.

In Swarm robotics, the key issues are communication between units, and cooperative behavior. The capability of individual units nodes not much matter; it is the strength in numbers. Ants and other social insects such as termites, wasps, and bees, are models for robot swarm behavior. Self-organizing behavior emerges from decentralized systems that interact with members of the group, and the environment. Swarm intelligence is an emerging field, and swarm robotics is in its infancy. Co-operative behavior, enabled by software and intra-unit communications has been demonstrated.

A Swarm exploration of the asteroid belt was proposed by Curtis, et al, in 2003. They baselined about 1,000 units. They defined` 8-10 types of *Workers,* each with specific capabilities. Units assigned to swarm cohesion and communication they term *Messengers.* There were also Rulers, who function in a managerial role. Cubesats are not specifically mentioned, but the approach is certainly feasible.

They postulate target selection according to mission goals, but also mention that mission goals change as data is collected at the site. The concept of multiple spacecraft coming together to form virtual instruments is discussed. Here, we might have simultaneous observations from multiple points.

The Operational Concept involves teams that produce data and some higher level products, which are communicated to Messengers, and archived. The Rulers oversee data flow. When a sufficient amount is collected, a Messenger will be dispatched to carry it back (Today, this could be accomplished with radio or laser link).

Swarms can be implemented with central control (Master-slaves) or distributed control (multi-master). The author worked with several student groups to design a mission to deploy a Cubesat Swarm to the Asteroid Belt and the multitude of moons at Jupiter and Saturn.

Not Just a Sat

Cubesats started out as inexpensive satellites, that could be afforded by schools. That has worked out. Well. Scientists looked at the capability and costs, and decided that proof-of-conecpt missions could be Cubesat-based, and much less expensive than traditional approaches.

You can also consider a Cubesat as a generic observation system, with various sensors, linked to an onboard computer, radio system, solar power, etc. There is no particular reason that they have to go to orbit. You can take that generic architecture, and bolt it on another mobility platform with track, wheels, or leg, or wings or rotors, or something that floats, or goes underwater. The basic principles are the same, but changes have to be made, of course, to address the specific operating environment. One interesting application is the Antarctic Meteor Finder. There is a vast expanse of white, with some small black rocks of great interest scattered about. For some reason, a lot of meteorites hit Antacrtica.

Co-operating units operating in different environments can also be used. For example, a drone can fly ahead of a ground rover, to spot crevasses and other hazards, as well as areas of interest. A tracked vehicle can drill through the ice sheet, to deploy a submarine craft to study the ice-water interface. They can examine the interior of nuclear power plants, with proper radiation hardening, and are cheap enough to be considered expendable.

The convergence of multiple technologies, along with the rapidly decreasing cost of highly capable systems, has led to University, high school, grade school, and even individual efforts. We are at the point with CubeSats where an individual can reasonably consider having his own payload launched into Earth orbit.

Cubesat architectures can ascend to the edge of space on balloon platforms. Again, ignore the "Sat' part of think of a sensor platform with a power system, computer, battery, radio, etc.

Trending

Expect to see increasing use of swarms of smallsats in place of single, large spacecraft explorers. As the technology advances, the units will become more capable. Cubesats represent a paradigm shift in orbital payloads. NASA itself uses cubesats as cost-effective small and relatively inexpensive explorers.

Onboard Software

Flight s/w is a special case of embedded software. As such, it is generally more difficult to design, implement, and test. It must be treated carefully, because most of the Cubesat functionality will rely on software, and the mission success will be directly related to software.

Flight Software can be proprietary or Open Source, but almost all Cubesat onboard software is open source.

FSW has several distinguishing characteristics:

- There are no direct user interfaces such as monitor and keyboard. All interactions are through uplink and downlink.

- It interfaces with numerous hardware devices such as science instruments and sensors

- It executes on radiation-hardened processors and microcontrollers that are relatively slow and memory-limited.

- It performs real-time processing. It must satisfy numerous timing constraints (timed commands, periodic deadlines, async event response). Being late = being wrong.

- Besides attitude determination and control, the onboard embedded systems has a variety of housekeeping tasks to attend to.

NASA's Core Flight Executive, and Core Flight Software

The Core Flight Executive, from the Flight Software Branch at NASA/GSFC, is an open source operating system framework. The executive is a set of mission independent reusable software services and an operating environment. Within this architecture, various mission-specific applications can be hosted. The cFE focuses on the commonality of flight software. The Core Flight System (CFS) supplies libraries and applications. Much flight software legacy went into the concept of the cFE. It has gotten traction within the Goddard community, and is in use on many flight projects, simulators, and test beds (FlatSats) at multiple NASA centers, as well as functioning in on-orbit Cubesat. The second application using the Goddard software was a drone project.

The cFE presents a layered architecture, starting with the bootstrap process, and including a real time operating system. At this level, a board support package is needed for the particular hardware in use. Many of these have been developed. At the OS abstraction level, a Platform support package is included. The cFE core comes next, with cFE libraries and specific mission libraries. Ap's habituate the 5_{th}, or upper layer. The cFE strives to provide a platform and project independent run time environment.

The boot process involves software to get things going after power-on, and is contained in non-volatile memory. cFE has boot loaders for the ARM, and other popular flight architectures. The real time operating systems can be any of a number of different open source or proprietary products, VxWorks and RTEMS for example. This layer provides interrupt handling, a scheduler, a file system, and interprocess communication.

The Platform Support Package is an abstraction layer that allows the cFE to run a particular RTOS on a particular hardware platform. There is a PSP for desktop pc's for the cFE. The cFE Core includes a set of re-usable, mission independent services. It presents a standardized Application Program Interface (API) to the programmer. A software bus architecture is provided for messaging between applications.

The Event services at the core level provides an interface to send asynchronous messages, telemetry. The cFE also provides time services.

Aps include a Health and Safety Ap with a watchdog. A housekeeping AP for messages with the ground, data storage and file manager aps, a memory checker, a stored command processor, a scheduler, a check-summer, and a memory manager. Aps can be developed and added to the library with ease.

A recent NASA/GSFC Cubesat project uses a FPGA-based system on a chip architecture with Linux and the cFE. CFE and its associated cFS are available as an architecture for Cubesats in general. The cFE has been released into the World-Wide Open Source community, and has found many applications outside of NASA.

NASA's Software Architecture Review Board reviewed the cFE in 2011. They found it a well thought-out product that definitely met NASA's needs. It was also seen to have the potential of becoming a dominant flight software architectural framework. The technology was seen to be mature.

The cFS is the core flight software, a series of aps for generally useful tasks onboard the spacecraft. The cFS is a platform and project independent reusable software framework and set of reusable applications. This framework is used as the basis for the flight software for satellite data systems and instruments, but can be used on other embedded systems in general. More information on the cFS can be found at http://cfs.gsfc.nasa.gov/OSAL

The OS Abstraction Layer (OSAL) project is a small software library that isolates the embedded software from the real time operating system. The OSAL provides an Application Program Interface (API) to an abstract real time operating system. This provides a way to develop one set of embedded application code that is independent of the operating system being used. It is a form of middleware.

Onboard File Systems

A file system provides a way to organize data in a standard format. An embedded system, like a digital camera, can store and organize its data (photos) and exchange the data directly with a computer. The file system stores the data, and metadata (data about the data) such as date, time, permissions, etc. Some operating systems support multiple file systems.

The important thing about a file systems for embedded systems is, don't reinvent the wheel! There are many good file systems out there, and the provide a compatibility across platforms. Most are based on the original disk operating system (dos) model.

The legacy DOS file structure is built upon linked lists. The directory file contains lists of files and information about them. It uses a 32-byte entry per file, containing the file name, extension, attributes, date and time, and the starting location of the file on disk or flash. Linux uses a similar approach.

Electrical Power

The electrical power for a Cubesat comes from batterys charged by solar panels. A charge regulator circuit makes sure the batterys are not overcharged. In telemetry, the Cubesat can include its state of charge.

How do you send data to and receive data from a Cubesat?

If your Cubesat probably has a Raspberry Pi or similar flight computer. It may also have one of several available off-the-shelf transceiver units. You may use a Software Defined Radio module running on the main computer. You may use a shared uplink/downlink service, such as COSMOS. In the latter case, you may want to add your own secure Virtual Private Network.

Ground Station

Don't have a 10 meter dish in your backyard? No problem. There was a lot of synergy between the Amateur Radio Operators and their Amsats, and the early Cubesats.

Now, an opensource ground station can be built with ease. The author watched a couple of students do it over a couple of days. Making the antenna is fairly easy, and a Raspberry Pi computer board hosts a Software-defined radio function. There are several of these available. One good example is the open source SATNOGS project.

A Control Center for a Cubesat mission only needs a laptop. The Open Source COSMOS software product from Ball Brothers Aerospace gives the full functionality of a million dollar, dedicated satellite center. Again, with students in a Cubesat course over the summer, the author was impressed. Those lazy students didn't want to walk from the dorms next door to the control center. So, they hacked COSMOS, first, to text them in the event of yellow or red limits being exceeded. Then they went on to integrate COSMOS with the Apache web server, putting the data on the internet. At this point, you don't even need your own control center and ground station, you can "rent" one as a service. So, you could in theory (and its probably been done) operate your Cubesat and get your data on your phone. Some NASA missions operate this way.

One control center can operate multiple satellites, as long as they have unique identifying numbers. Unless you're in geostationary orbit, you only get data 10 minutes out of 90. There is a growing network of opensource amateur ground stations, so it should be possible to get data 24x7.

NASA is building a "virtual telescope" using two Cubesats. They are imaging the Sun, in a mission called "Cubesat Astronomy by NASA and Yonsei using Virtual Telescope alignment experiment," CanyVAL-X. Two spacecrafts, flying in coordination and aligned with the Sun are being used. The one closest to the sun blocks the solar sphere, allowing the second spacecraft to image the outer regions of the solar atmosphere. The spacecraft are named Tom and Jerry. Jerry is smaller (1U), and Tom (2U) is between Jerry and the Sun. They both have solar sensors, Tom uses a camera to look at Jerry's laser beacons to keep alignment. The spacecraft are separated by 10 meters. This is an early proof-of-concept mission that was launched in 2017. The mission cost around $1 million which is a lot for 2 Cubesats, but a drop in the bucket for a full sized spacecraft.

The Cubesat Space Protocol

For Cubesats, which until recently were confined to Earth orbit, we would prefer a communications implementation in Open Source. There are implementations of various communications protocols available in Open Source format for the popular Raspberry Pi architecture, and similar units, even the 8-bit AVR. There is a specific Cubesat Space security protocol, based on the same layers as TCP/IP. It does include support for encrypted packets.

Spacecraft Cybersecurity Policy for non-DoD payloads is controlled by the sponsors, NASA, NOAA (for remote sensing missions), and the FAA for private missions. Each of these entity's has their own policy's. Commercial payloads fall under the National Commercial and Space Programs Act, with licensing authority given to NOAA for imager missions. NASA has no current requirement to encrypt. Private, non-imaging concerns operate under the lack of co-ordinated public policy in this matter.

First Interplanetary Cubesats

This section discusses the first Cubesats to leave Earth orbit. NASA's Planetary Cubesat Science Institute (PCSI) is focusing on developing Cubesat approaches for high priority science. In August of 2016, they held a symposium at Goddard Space Flight Center, getting key people together to exchange ideas. Projects include the Hydrogen Albedo Lunar Orbiter (HALO) and the Primitive Object Volatile Explorer (Prove), which did a close fly-by of the comet 46P at its perihelion in December 2018, sampling the volatiles that are boiling off.

Once we leave the vicinity of our home planet, conditions deteriorate quickly. The major issue is radiation, since we are outside of the trapped radiation belts, which provide some protection. This is a major challenge, but there are many known ways to mitigate this problem. Then, there is the thermal problem. We're going somewhere that's hotter (sun-ward), or colder. A big issue is the mission duration. It takes years to get to some of the outer planets, and even if the system is powered off, there can be events that will cause it to not wake up. Missions outside the rather friendly environment of near-Earth face additional challenges that must be addressed.

At the same time, communications becomes more difficult, and achievable data rates go down. The spacecraft might find itself on the other side of the Sun, from Earth, and communications would not be possible. JPL is exploring laser communication links for long distances.

Cubesats, being small, have constraints on power generation and storage, fuel storage (if any) and communications.

These are all solvable problems, but require additional engineering analysis.

Communications between planets in our solar system involves long

24

distances, and significant delay. New protocols were needed to address the long delay times, and error sources.

A concept called the Interplanetary Internet uses a store-and-forward node in orbit around a planet (initially, Mars) that burst-transmits data back to Earth during available communications windows. At certain times, when the geometry is right, the Mars bound traffic might encounter significant interference. Mars surface craft communicate to Orbiters, which relay the transmissions to Earth. This allows for a lower wattage transmitter on the surface vehicle. Mars does not (yet) have the full infrastructure that is currently in place around the Earth – a network of navigation, weather, and communications satellites.

For satellites in near Earth orbit, protocols based on the cellular terrestrial network can be used, because the delays are small. In fact, the International Space Station is a node on the Internet. By the time you get to the moon, it takes about a second and a quarter for electromagnetic energy to traverse the distance. Delay tolerant protocols developed for mobile terrestrial communication were used, but break down in very long delay situations.

In 2015, the Planetary Society's LightSail-1 successfully deployed its solar sail. This was done in Earth orbit. Planned follow-on projects LightSail-2, 3, and 4 would follow. Lightsail-2 was launched with 32 square meters of sail, and advanced guidance electronics. It is a 3U Cubesat. It will deploy its sail at 800km. LightSat's 3 and 4 will be more than technology demonstration, with the 4th unit heading to the L1 Lagrange point, to provide earlier warning of Solar geomagnetic storms. Other systems have been proposed with continuous low-thrust ion engines. All these approaches require specific new trajectory designs. There is increasing effort in applying non-linear, non-Keplerian orbits.

Marco

The MARCO mission, in 2018, had dual 6-U Cubesat fly-alongs, that separated after launch, and continue to Mars along with the primary payload, a rover. The Cubesats serve as a real-time communications relay with Earth during the critical descent and landing phase of the rover. The Rover talks to the Cubesat relays over an 8kbps UHF link, and the Cubesats send this to Earth over an 8kbps X-band link to the DSN. The Cubesat's X-band antenna is a large flat panel. MARCO successfully returned images.

Dellingr

Dellinger is a Cubesat bus design from NASA/GSFC. The concept was to design and implement a cost-efficient and reliable satellite bus for science payloads. It was carried to the International Space Station, and deployed from there following checkout. The satellite has a payload of an Ion-Neutral Mass Spectrometer (INMS), and dual magnetometers.

The same Dellinger bus design has also flown with a different science instrument set. It is called PetitSat, Plasma Enhancements in the Ionosphere and Thermosphere Satellite.

Capstone

Capstone, an acronym for Cislinur Autonomous, Positioning System Technology, is scheduled to be launched from the Wallops Flight Facility in the Fall of 2021. It is a 12U cubesat commercial platform. It will spend three months getting to the moon, and is scheduled to do 6 months of observation. It will collect data on orbital stability that will affect the planned Lunar Gateway project.

Hera

The ESA Hera mission involves a tag-along pair of 6U Cubesats observing the impact of a NASA impactor into the moon of a binary asteroid system called Didymos. The NASA part of the mission involves the DART impactor. The main spacecraft, Hera, serves as the Mothership. One Cubesat is known as APX, the Asteroid Prospector Explorer. I will conduct spectral analysis and magnetic readings at close-up distances from the asteroid's surface.

The second cubesat is called Juventas, and will do gravity field measurements to explore the internal structure of the asteroid.

LUCE

LUCE is an 12-U ESA Cubesat whose mission involves a ride-share to the moon. It is part of a mission called SOLVE, small spacecraft for near lunar environment exploration. A lunar orbiter provides transportation, and data relay services. The orbit will be gradually lowered, to do a soft surface landing. This is to be a demonstration of a soft landing on the Moon by a Cubesat.

M-ARGO

ESA's M-ARGO is an 12-U ESA Cubesat payload planned for 2024. It will study a Near Earth Object. It is a stand-along mission. The name is an acronym, *Miniaturized Asteroid Remote Geophysical Observer*. It will be launched in a ride-share arrangement. It uses an electric propulsion system to travel to its target, which is an asteroid. It will hopefully be able to visit multiple targets. It will use a high-gain antenna to communicate with ESA's Estrack ground stations, as well as the Italian Sardinia Radio Telescope facility. It will take it three years to cover the 150 million miles of its journey.

Asteria

Asteia is a JPL 6U Cubesat-based astrophysics mission, launched to the International Space Station in 2017. It was deployed from there is November, 2017. It was JPL's first (but certainly not last) Cubesat. It achieved its primary mission requirements, to track selected guide stars for long periods. It's communication was lost in February of 2020.

Brazil's Tancredo-1

This interesting STEM project by middle school students was sent to the Space Station at the end of 2016, and deployed into space from the Japanese module. It was built from a kit from IOS, in California. It has the support of the Brazilian Space Agency, The National Institute for Space Research, and Unesco. It featured a pre-recorded message from one of the school team members, that was broadcast from orbit.

ZACUBE-1

From the Cape Peninsula University of Technology in South Africa is Africa's first Cubesat. It was built as a project of the South African National Space Agency, with assistance from the French. It was launched on a Russian vehicle in 2013, and is collecting data on space weather. The project has resulted not only in on-orbit success, but 22 Master's degrees, 10 conference papers, and 3 journal papers. The project also spun off the African Space Innovation Center, with a Research Chair.

Radiation Hardness Issues for Space Flight Applications

A complete discussion of the physics of radiation damage to semiconductors is beyond the scope of this book However, an overview of the subject is presented. The tolerance of semiconductor devices to radiation must be examined in the light of their damage susceptibility. The problems fall into two broad categories, those caused by cumulative dose, and those transient events caused by asynchronous very energetic particles, such as those experienced during a period of intense solar flare activity. The unit of absorbed dose of radiation is the *rad*, representing the absorption of 100 ergs of energy per gram of material. A kilo-rad is one thousand rads. At 10k rad, death in humans is almost instantaneous. One hundred kilo-rad is typical in the vicinity of Jupiter's radiation belts. Ten to twenty kilo-rad is typical for spacecraft in low Earth orbit, but the number depends on how much time the spacecraft spends outside the Van Allen belts, which act as a shield by trapping energetic particles.

Absorbed radiation can cause temporary or permanent changes in the semiconductor material. Usually neutrons, being uncharged, do minimal damage, but energetic protons and electrons cause lattice or ionization damage in the material, and resultant parametric changes. For example, the leakage current can increase, or bit states can change. Certain technologies and manufacturing processes are known to produce devices that are less susceptible to damage than others. More expensive substrate materials such as diamond or sapphire help to make the device more tolerant of radiation, but much more expensive.

Radiation tolerance of 100 kilo-rad is usually more than adequate for low Earth orbit (LEO) missions that spend most of their life below the shielding of the Van Allen belts. For Polar missions, a higher total dose is expected, from 100k to 1 mega-rad per year. For synchronous, equatorial orbits, that are used by many communication satellites, and some weather satellites, the expected dose is several kilo-rad per year. Finally, for planetary missions to Venus, Mars, Jupiter, Saturn, and beyond, requirements that are even more stringent must be met. For one thing, the missions usually are unique, and the cost of failure is high. For missions towards the sun, the higher fluence of solar radiation must be taken into account. The larger outer planets, such as Jupiter and Saturn, have their own large radiation belts around them as well.

Cumulative radiation dose causes a charge trapping in the semi-conductor oxide layers, which manifests as a parametric change in the devices. Total dose effects are a function of the dose rate, and annealing of the device may occur, especially at elevated temperatures. Annealing refers to the self-healing of radiation induced defects. This can take minutes to months, and is not applicable for lattice damage. The internal memory or registers of the cpu are the most susceptible area of the chip, and are usually deactivated for operations in a radiation environment. The gross indication of radiation damage is the increased power consumption of the device, and one researcher reported a doubling of the power consumption at failure. In addition, failed devices would operate at a lower clock rate, leading to speculation that a key timing parameter was being effected in this case.

Single event upsets (seu's) are the response of the device to direct high energy isotropic flux, such as cosmic rays, or the secondary effects of high energy particles colliding with other matter (such as shielding). Large transient currents may result, causing changes in logic state (bit flips), unforeseen operation, device latch-up, or burnout. The transient currents can be monitored as an indicator of the onset of SEU problems. After SEU, the results on the operation of the processor are unpredictable. Mitigation of problems caused by SEU's involves self-test, memory scrubbing, and forced resets.

The LET (linear energy transfer) is a measure of the incoming particles' delivery of ionizing energy to the device. Latch-up refers to the inadvertent operation of a parasitic SCR (silicon control rectifier), triggered by ionizing radiation. In the area of latch-up, the chip can be made inherently hard due to use of the Epitaxial process for fabrication of the base layer. Even the use of an Epitaxial layer does not guarantee complete freedom from latch-up, however. The next step generally involves a silicon on insulator (SOI) or Silicon on Sapphire (SOS) approach, where the substrate is totally insulated, and latch-ups are not possible. This is an expensive approach,

In some cases, shielding is effective, because even a few millimeters of aluminum can stop electrons and protons. However, with highly energetic or massive particles (such as alpha particles, helium nuclei), shielding can be counter-productive. When the atoms in the shielding are hit by an energetic particle, a cascade of lower energy, lower mass particles results. These can cause as much or more damage than the original source particle.

Cumulative dose and single events

The more radiation that the equipment gets, in low doses for a long time, or in high doses for a shorter time, the greater the probability of damage. The Total Ionization Dose (TID) accumulates over time, and actually displaces the semiconductor lattice structure. It causes shifts in the threshold voltage device, and noticeable increased current draw. The damage can become permanent. TID isn't the major concern, as devices become smaller, and the oxide gates become thinner, as technology advances. The higher the voltage, though, the more problematic the effect can be. Analog to digital converters can experience conversions shifts.

These events are caused by high energy particles, usually protons, that disrupt and damage the semiconductor lattice. The effects can be upsets (bit changes) or latch-ups (bit stuck). The damage can "heal" itself, but its often permanent. Most of the problems are caused by energetic solar protons, although galactic cosmic rays are also an issue. Solar activity varies, but is now monitored by sentinel spacecraft, and periods of intensive solar radiation and particle flux can be predicted. Although the Sun is only 8 light minutes away from Earth, the energetic particles travel much slower than light, and we have several days warning. During periods of intense solar activity, Coronal Mass Ejection (CME) events can send massive streams of charged particles outward. These hit the Earth's magnetic field and create a bow wave. The Aurora Borealis or Northern Lights are one manifestation of incoming charged particles hitting the upper reaches of the ionosphere.

Cosmic rays, particles and electromagnetic radiation, are omni-directional, and come from extra-solar sources. Most of them, 85%, are protons, with gamma rays and x-rays thrown in the mix. Energy levels range to 10^6 to 10^8 electron volts (eV). These are mostly filtered out by Earth's atmosphere. There is no such mechanism on the Moon, where they reach and interact with the surface material. Solar flux energy's range to several Billion electron volts (Gev).

The effects of radiation on silicon circuits can be mitigated by redundancy, the use of specifically radiation hardened parts, Error

Detection and Correction (EDAC) circuitry, and scrubbing techniques. Bipolar technology chips can withstand radiation better than CMOS technology chips, at the cost of greatly increased power consumption. Shielding techniques are also applied. Radiation hardened parts are much more expensive than standard parts.

EDAC can be done with hardware or software, but always carries a cost in time and complexity. A longer word than needed for the data item allows for the inclusion of error detecting and correcting codes. The simplest scheme is a parity bit, which can detect single bit (or an odd number of errors, but can't correct anything. EDAC is applied to memory and I/O, particularly the uplink and downlink.

Single Event Upsets (SEU) are instantaneous events, caused by highly energetic particles such as Cosmic Rays. This causes momentary bit flips, but is generally not cumulative. Some events may require a reset to affect recovery of state.
The effects of radiation on silicon circuits can be mitigated by redundancy, the use of specifically radiation hardened parts, Error Detection and Correction (EDAC) circuitry, and scrubbing techniques. Hardened chips are produced on special insulating substrates such as Sapphire. Bipolar technology chips can withstand radiation better than CMOS technology chips, at the cost of greatly increased power consumption. Shielding techniques are also applied. In error detection and correction techniques, special encoding of the stored information provides a protection against flipped bits, at the cost of additional bits to store. Redundancy can also be applied at the device or box level, with the popular Triple Modular Redundancy (TMR) technique triplicating everything, and based on the assumption that the probability of a double failure is less than that of a single failure. Watchdog timers are used to reset systems unless they are themselves reset by the software. Of course, the watchdog timer circuitry is also susceptible to failure.

One concept that is easily implemented, and addresses the

radiation damage issue, is called Rad Hard software. This is a series of software routines that run in the background on the flight computer, and check for the signs of radiation damage. The biggest indicator is an increase in current draw. The flight cpu must monitor and trend it's current draw, and take critical action such as a reboot if it deems necessary. The Rad Hard software is a variation on self-check routines, but with the ability to take action if needed. We can keep tabs on memory by conducting CRC (cyclic redundancy checks), and one approach to mitigating damage to semiconductor memory is "scrubbing," where we read and write back each memory locations (being careful not to interfere with ongoing operations). This can be done by a background task that is the lowest priority in the system. Watchdog timers are also useful in getting out of a situation such as the Priority Inversion, or just a radiation-induced bit flip. There should be a pre-defined safe mode for the computer as well. Key state data from just before the fault should be telemetered to the control center. Unused portions of memory can be filled with bit patterns that can be monitored for changes. We must be certain that all of the unused interrupt vectors point to a safe area in the code. There is a lot of creative work to be done in this area.

Afterword

Not all space missions can be done with a Cubesat, but the little guys represent a paradigm shift in our way of thinking about these missions. Veteran Rocket Scientists reconsidered the implementation of a mission, that can be done for far less cost. Grade school kids can imagine their cubesat on the ISS. In the far reaches of the solar system, a swarm of cubesats can do in situ science on unexplored worlds.

We will see more and more schools, university's, country's, private companys, and individuals launching cubesats. Some companys are launching them in the hundreds, for example, to provide 5G wifi from orbit. Its going to get crowded up there.

Glossary

1553 – Military standard data bus, serial, 1 Mbps.

1U – one unit for a Cubesat, 10 x 10 x 10 cm.

3U – three units for a Cubesat

6u – 6 units in size, where a unit is defined by dimensions and
   weight.

802.11 – a radio frequency wireless data communications standard.

AACS – (JPL) Attitude and articulation control system.

ACE – attitude control electronics

Actuator – device which converts a control signal to a mechanical
   action.

A/D, ADC – analog to digital converter

AEB - Agência Espacial Brasileira

AFB – Air Force Base.

AGC – Automated guidance and control.

AIAA – American Institute of Aeronautics and Astronautics.

AIST – NASA GSFC Advanced Information System Technology.

ALU – arithmetic logic unit.

AmSat – Amateur Satellite. Favored by Ham Radio operators as
   communication relays.

Analog – concerned with continuous values.

ANSI – American National Standards Institute

Android – an operating system based on Gnu-Linux, popular for
   smart phones and tablet computers.

Antares – Space launch vehicle, compatible with Cubesats, by
Orbital/ATK (U.S.)

AP – application programs.

API – application program interface; specification for software
   modules to communicate.

APL – Applied Physics Laboratory, of the Johns Hopkins
   University.

Apm – antenna pointing mechanism

Apollo – US manned lunar program.

Amsat – amateur satellite, usually a relay for radio amateurs.

Arduino – a small, inexpensive microcontroller architecture.

ArduSat – Arduino-based Cubesat

Arinc – Aeronautical Radio, Inc. commercial company supporting transportation, and providing standards for avionics.

ARM – Acorn RISC machine; a 32-bit architecture with wide application in embedded systems.

ARPA – (U. S.) Advanced Research Projects Agency.

ArpaNet – Advanced Research Projects Agency (U.S.), first packet switched network, 1968.

ASIC – application specific integrated circuit.

ASIN – Amazon Standard Inventory Number.

Asteria (JPL) Arcsecond Space Telescope Enabling Research in Astrophysics, a Cubesat mission.

async – non synchronized

ATAC – Applied Technologies Advanced Computer.

ATP – authority to proceed

AU – astronomical unit. Roughly 93 million miles, the mean distance between Earth and Sun,

BAE – British Aerospace.

Baud – symbol rate; may or may not be the same as bit rate.

BCD – binary coded decimal. 4-bit entity used to represent 10 different decimal digits; with 6 spare states.

Beowolf – a cluster of commodity computers; multiprocessor, using Linux.

Big-endian – data format with the most significant bit or byte at the lowest address, or transmitted first.

Binary – using base 2 arithmetic for number representation.

BIST – built-in self test.

Bit – binary variable, value of 1 or 0.

Boolean – a data type with two values; an operation on these data types; named after George Boole, mid-19th century inventor of Boolean algebra.

Bootloader – initial program run after power-on or reset. Gets the computer up & going.

Bootstrap – a startup or reset process that proceeds without external intervention.

BSD – Berkeley Software Distribution version of the Bell Labs Unix operating system.

BP - bundle protocol, for dealing with errors and disconnects.

BSP – board support package. Customization Software and device drivers.

Buffer – a temporary holding location for data.

Bug – an error in a program or device.

Bus – an electrical connection between 2 or more units; the engineering part of the spacecraft.

byte – a collection of 8 bits

C – programming language from Bell Labs, circa 1972.

cache – temporary storage between cpu and main memory.

Cache coherency – process to keep the contents of multiple caches consistent,

CalPoly – California Polytechnic State University, San Luis Obispo, CA.

CAN - controller area network bus.

CAPS – Cislunar Autonomous Positioning System.

CCSDS – Consultive Committee on Space Data Systems.

CDR – critical design review

C&DH – Command and Data Handling

CDFP CCSDS File Delivery Protocol

cFE – Core Flight Executive – NASA GSFC reusable flight software.

CFS – Core Flight System – NASA GSFC reusable flight software.

Chip – integrated circuit component.

Clock – periodic timing signal to control and synchronize operations.

CME – Coronal Mass Ejection. Solar storm.

CMOS – complementary metal oxide semiconductor; a technology using both positive and negative semiconductors to achieve low power operation.

CogE – cognizant engineer for a particular discipline; go-to guy; specialist.

Complement – in binary logic, the opposite state.

Compilation – software process to translate source code to assembly or machine code (or error codes).

Configware – equivalent of software for FPGA architectures;

configuration information.

Constellation – a grouping of satellites.

Control Flow – computer architecture involving directed flow through the program; data dependent paths are allowed.

COP – computer operating properly.

Coprocessor – another processor to supplement the operations of the main processor. Used for floating point, video, etc. Usually relies on the main processor for instruction fetch; and control.

Cordic – Coordinate Rotation Digital Computer – to calculate hyberbolic and trig functions.

Cots – commercial, off the shelf

CPU – central processing unit

CRC – cyclic redundancy code – error detection and correction mechanism.

CSP – Cubesat Space Protocol

CubeRover – a small planetary rover following the Cubesat modular architecture.

Cubesat – small inexpensive satellite for colleges, high schools, and individuals.

D/A – digital to analog conversion.

DAC – digital to analog converter.

Daemon – in multitasking, a program that runs in the background.

DARPA – Defense advanced research projects agency. (U. S.)

Dataflow – computer architecture where a changing value forces recalculation of dependent values.

Datagram – message on a packet switched network; the delivery, arrival time, and order of arrival are not guaranteed.

dc – direct current.

D-cache – data cache.

DDR – dual data rate memory.

Deadlock – a situation in which two or more competing actions are each waiting for the other to finish, and thus neither ever does.

DCE – data communications equipment; interface to the network.

Deadly embrace – a deadlock situation in which 2 processes are each waiting for the other to finish.

Denorm – in floating point representation, a non-zero number with a magnitude less than the smallest normal number.

Device driver – specific software to interface a peripheral to the operating system.

Digital – using discrete values for representation of states or numbers.

Dirty bit – used to signal that the contents of a cache have changed.

Discrete – single bit signal.

DMA – direct memory access.

Dnepr – Russian space launch system compatible with Cubesats.

DOD – (U. S.) Department of Defense.

DOE – (U. S.) Department of Energy.

DOF – degrees of freedom.

Downlink – from space to earth.

Dram – dynamic random access memory.

DSN (NASA) Deep Space network.`

DSP – digital signal processing/processor.

DTE – data terminal equipment; communicates with the DCE to get to the network.

DTN – delay tolerant networks.

DUT – device under test.

ECC – error correcting code

EDAC – error detecting and correction circuitry.

EDS – Electronic Data Sheets

EGSE – electrical ground support equipment

EIA – Electronics Industry Association.

ElaNa – Educational Launch of Nanosatellites.

ELV – expendable launch vehicle.

Embedded system – a computer systems with limited human interfaces and performing specific tasks. Usually part of a larger system.

EMC – electromagnetic compatibility.

EMI – electromagnetic interference.

EOL – end of life.

EOS – Earth Observation spacecraft.

Ephemeris – orbital position data.

Epitaxial – in semiconductors, have a crystalline overlayer with a well-defined orientation.

EPS – electrical power subsystem.

ESA – European Space Organization.

ESRO – European Space Research Organization

ESTO – NASA/GSFC – Earth Science Technology Office.

Ethernet – networking protocol, IEEE 802.3

ev – electron volt, unit of energy

EVA – extra-vehicular activity.

Exception – interrupt due to internal events, such as overflow.

EXPRESS racks – on the ISS, EXpedite the PRocessing of Experiments for Space Station Racks

FAA – (U S.) Federal Aviation Administration.

Fail-safe – a system designed to do no harm in the event of failure.

Falcon – launch vehicle from SpaceX.

FCC – (U.S.) Federal Communications Commission.

FDC – fault detection and correction.

Femtosatellites - smaller than a Cubesat, 3.5 cm on a side.

Firewire – IEEE-1394 standard for serial communication.

Firmware – code contained in a non-volatile memory.

Fixed point – computer numeric format with a fixed number of digits or bits, and a fixed radix point.

Flag – a binary state variable.

Flash – non-volatile memory

Flatsat – prototyping and test setup, laid out on a bench for easy access.

FlightLinux – NASA Research Program for Open Source code in space.

Floating point – computer numeric format for real numbers; has significant digits and an exponent.

FPGA – field programmable gate array.

FPU – floating point unit, an ALU for floating point numbers.

Fram – ferromagnetic RAM; a non-volatile memory technology

FRR – Flight Readiness Review

FSW – flight software.

FTP – file transfer protocol

Gbyte – 10^9 bytes.

GEO – geosynchronous orbit.

GeV – billion (10^9) electron volts.

GMSEC – Goddard Mission Services Evolution Center.

GNC – guidance, navigation, and control.

Gnu – recursive acronym, gnu is not unix.

GPIO – general purpose I/O.

GPL – gnu public license used for free software; referred to as the "copyleft."

GPS – Global Positioning system – Navigation satellites.

GPU – graphics processing unit. ALU for graphics data.

GSFC – Goddard Space Flight Center, Greenbelt, MD.

Gyro – (gyroscope) a sensor to measure rotation.

Half-duplex – communications in two directions, but not simultaneously.

HAL/S – computer language.

Handshake – co-ordination mechanism.

HDL – hardware description language

Hertz – cycles per second.

Hexadecimal – base 16 number representation.

Hi-rel – high reliability

HPCC – High Performance Computing and Communications.

Hypervisor – virtual machine manager. Can manage multiple operating systems.

I2C – a serial communications protocol.

IARU – International Amateur Radio Union

I-cache – Instruction cache.

ICD – interface control document.

IC&DH – Instrument Command & Data Handling.

IEEE – Institute of Electrical and Electronic engineers

IEEE-754 – standard for floating point representation and calculation.

IIC – inter-integrated circuit (I/O).

IMU – inertial measurement unit.

INPE Instituto Nacional de Pesquisas Espaciais (Brazilian National Institute for Space Research)

Integer – the natural numbers, zero, and the negatives of the natural numbers.

Interrupt – an asynchronous event to signal a need for attention (example: the phone rings).

Interrupt vector – entry in a table pointing to an interrupt service routine; indexed by interrupt number.

IP – intellectual property; Internet protocol.

IP core – IP describing a chip design that can be licensed to be used in an FPGA or ASIC.

IP-in-Space – Internet Protocol in Space.

IR – infrared, 1-400 terahertz. Perceived as heat.

IRAD – Independent Research & Development.

ISA – instruction set architecture, the software description of the computer.

ISO – International Standards Organization.

ISR – interrupt service routine, a subroutine that handles a particular interrupt event.

ISRO – Indian Space Research Organization

ISS – International Space Station

I&T – integration & test

ITAR – International Trafficking in Arms Regulations (US Dept. of State)

ITU – International Telecommunications Union

IV&V – Independent validation and verification.

JEM – Japanese Experiment Module, on the ISS.

JHU – Johns Hopkins University.

JPL – Jet Propulsion Laboratory

JSC – Johnson Space Center, Houston, Texas.

JTAG – Joint Test Action Group; industry group that lead to IEEE 1149.1, Standard Test Access Port and Boundary-Scan Architecture.

JWST – James Webb Space Telescope – follow on to Hubble.

Kbps – kilo (103) bits per second.

Kernel – main portion of the operating system. Interface between the applications and the hardware.

Kg – kilogram.

kHz – kilo (103) hertz

KVA – kilo volts amps – a measure of electrical power

Ku band – 12-18 Ghz radio

Lan – local area network, wired or wireless

LaRC – (NASA) Langley Research Center.

Latchup – condition in which a semiconductor device is stuck in one state.

Lbf – pounds-force.

LEO – low Earth orbit.

Let- Linear Energy Transfer

Lidar – optical radar.

Linux – open source operating system.

List – a data structure.

Little-endian – data format with the least significant bit or byte at the highest address, or transmitted last.

Logic operation – generally, negate, AND, OR, XOR, and their inverses.

Loop-unrolling – optimization of a loop for speed at the cost of space.

LRO – Lunar Reconnaissance Orbiter.

LRR – launch readiness review

LRU – least recently used; an algorithm for item replacement in a cache.

LSB – least significant bit or byte.

LSP – (NASA) launch services program, or launch services provider

LUT – look up table.

MARS – Mid-Atlantic Regional Spaceport (Wallops Flight Facility, Virginia)

Master-slave – control process with one element in charge. Master status may be exchanged among elements.

Mbps – mega (10^6) bits per second.

Mbyte – one million (10^6 or 2^{20}) bytes.

MEMS – Micro Electronic Mechanical System.

MESI – modified, exclusive, shared, invalid state of a cache coherency protocol.

MEV – million electron volts.

MEXEC - Multi-mission EXECutive(JPL), integrated, task-based, planning and execution library.

MHz – one million (10^6) Hertz

Microcontroller – monolithic cpu + memory + I/O.

Microkernel – operating system which is not monolithic, functions execute in user space.

Microprocessor – monolithic cpu.

Microsat – satellite with a mass between 10 and 100 kg.

Microsecond – 10-6 second.

Microkernel – operating system which is not monolithic. So functions execute in user space.

MIPS – millions of instructions per second.

MLI – multi-layer insulation.

MPA – multiple payload adapter for deploying multiple p-pod's

MPE – Maximum predicted environments.

mram – magnetorestrictive random access memory.

mSec – Millisecond; (10-3) second.

MMU – memory management unit; manned maneuvering unit.

MSB – most significant bit or byte.

Multiplex – combining signals on a communication channel by sampling.

Multicore – multiple processing cores on one substrate or chip; need not be identical.

Mutex – a software mechanism to provide mutual exclusion between tasks.

Nano – 10^{-9}

NanoRacks – a company providing a facility onboard the ISS to support Cubesats

NanoSat – small satellite with a mass between 1 and 10 kg.

NASA - National Aeronautics and Space Administration.

NDA – non-disclosure agreement; legal agreement protecting IP.

NEN – (NASA's) Near Earth Network.

NEO – Near Earth Object.

Nibble – 4 bits, ½ byte.

NIST – National Institute of Standards and Technology (US), previously, National Bureau of Standards.

NMI – non-maskable interrupt; cannot be ignored by the software.

Normalized number – in the proper format for floating point representation.

NRCSD - NanoRack CubeSat Deployer

NRE – non-recurring engineering; one-time costs for a project.

NSF – (U.S.) National Science Foundation.

NSR – non-space rated.

NTIA (U.S.) National Telecommunications and Information Administration

NUMA – non-uniform memory access for multiprocessors; local and global memory access protocol.

NVM – non-volatile memory.

NWS – (U.S.) National Weather Service

Nyquist rate – in communications, the minimum sampling rate, equal to twice the highest frequency in the signal.

OBC – on board computer

OBD – On-Board diagnostics.

OBP – On Board Processor

Off-the-shelf – commercially available; not custom.

OPAL - Orbiting Picosatellite Automatic Launcher

OpAmp – (Linear) operational amplifier; linear gain and isolation stage.

OpCode – encoded computer instruction.

Open source – methodology for hardware or software development with free distribution and access.

Operating system – software that controls the allocation of resources in a computer.

OSAL – operating system abstraction layer.

OSI – Open systems interconnect model for networking, from ISO.

Overflow - the result of an arithmetic operation exceeds the capacity of the destination.

Packet – a small container; a block of data on a network.

Paging – memory management technique using fixed size memory blocks.

Paradigm – a pattern or model

Paradigm shift – a change from one paradigm to another, disruptive or evolutionary.

Parallel – multiple operations or communication proceeding simultaneously.

Parity – an error detecting mechanism involving an extra check bit in the word.

PC-104 – standard for a board (90 x 96 mm), and a bus for

embedded use.

PCB – printed circuit board.

pci – personal computer interface (bus).

PCM – pulse code modulation.

PCSI – (NASA/GSFC) Planetary Cubesat Science Institute

PDR – preliminary design review.

Perseus-M 6U maritime surveillance cubesats

Peta - 10^{15} or 2^{50}

Phonesat – small satellite using a cell phone for onboard control and computation.

Picosat – small satellite with a mass between 0.1 and 1 kg.

Piezo – production of electricity by mechanical stress.

Pinout – mapping of signals to I/O pins of a device.

Pipeline – operations in serial, assembly-line fashion.

PiSat – a Cubesat architecture developed at NASA-GSFC, based on the Raspberry Pi architecture.

Pixel – picture element; smallest addressable element on a display or a sensor.

PLL – phase locked loop.

PocketQube – smaller than a Cubesat; 5 cm cubed, a mass of no more than 180 grams, and uses COTS components.

Poc – point of contact

POCC – payload Operations Control Center

POSIX – IEEE standard operating system.

PPF – payload processing facility

PPL – preferred parts list (NASA).

P-POD – Cubesat launch dispenser, Poly-Picosatellite Orbital Deployer.

Psia – pounds per square inch, absolute.

PSP – Platform Support Package.

PWM – pulse width modulation.

Python – programming language.

Quadrature encoder – an incremental rotary encoder providing rotational position information.

Queue – first in, first out data buffer structure; implemented in hardware or software.

Rad – unit of radiation exposure

Rad750 – A radiation hardened IBM PowerPC cpu.

Radix point – separates integer and fractional parts of a real number.

RAID – redundant array of inexpensive disks.

Ram – random access memory.

RBF – remove before flight.

Real-time – system that responds to events in a predictable, bounded time.

Register – temporary storage location for a data item.

Reset – signal and process that returns the hardware to a known, defined state.

RF – radio frequency

RFC – request for comment

RISC – reduced instruction set computer.

RHPPC – Rad-Hard Power PC.

RHS – rad-hard software

RISC – reduced instruction set computer.

Router – networking component for packets.

RS-232/422/423 – asynchronous and synchronous serial communication standards.

RT – remote terminal.

RTC – real time clock.

RTOS – real time operating system.

SAM – sequential access memory, like a magnetic tape.

Sandbox – an isolated and controlled environment to run untested or potentially malicious code.

SDR – software defined radio

SDRAM – synchronous dynamic random access memory.

Segmentation – dividing a network or memory into sections.

Semiconductor – material with electrical characteristics between conductors and insulators; basis of current technology processor, memory, and I/O devices, as well as sensors.

Semaphore – a binary signaling element among processes.

SD – secure digital (non-volatile memory card).

SDVF – Software Development and Validation Facility.

Sensor – a device that converts a physical observable quantity or event to a signal.

Serial – bit by bit.

SEU – single event upset (radiation induced error).

Servo – a control device with feedback.

SIMD – single instruction, multiple data (parallel processing)

Six-pack – a six U Cubesat, 10 x 20 x 30 cm.

SMP – symmetric multiprocessing.

Snoop – monitor packets in a network, or data in a cache.

SN – (NASA's) Space Network

SOA – safe operating area; also, state of the art.

SOC – system on a chip; also state-of-charge.

Socket – an end-point in communication across a network

Soft core – a hardware description language description of a cpu core.

Software – set of instructions and data to tell a computer what to do.

Smallsat – less than 180 kg.

SMP – symmetric multiprocessing.

Snoop – monitor packets in a network, or data in a cache.

Spacewire – high speed (160 Mbps) link.

SPI - Serial Peripheral Interface - a synchronous serial communication interface.

SRAM – static random access memory.

Stack – first in, last out data structure. Can be hardware or software.

Stack pointer – a reference pointer to the top of the stack.

STAR – self test and repair.

State machine – model of sequential processes.

STOL – system test oriented language, a scripting language for testing systems.

SAA – South Atlantic anomaly. High radiation zone.

SEB – single event burnout.

SEU – single event upset.

SEL – single event latchup.

Soc – state of charge; system on a chip.

Soft core – hardware description description language model of a logic core.

SOI – silicon on insulator

SoS – silicon on sapphire – an inherently radiation-hard technology

spi – serial peripheral interface

SpaceCube – an advanced FPGA-based flight computer.

SpaceWire – networking and interconnect standard.

Space-X – commercial space company.

SRAM – static random access memory.

Stack – first in, last out data structure. Can be hardware or software.

Stack pointer – a reference pointer to the top of the stack.

State machine – model of sequential processes.

SWD – serial wire debug.

Synchronous – using the same clock to coordinate operations.

System – a collection of interacting elements and relationships with a specific behavior.

System of Systems – a complex collection of systems with pooled resources.

Suitsat – old Russian spacesuit, instrumented with an 8-bit micro, and launched from the ISS.

Swarm – a collection of satellites that can operate cooperatively.

Sync – synchronize, synchronized.

TCP/IP – Transmission Control Protocol/Internet protocol.

TDRSS – Tracking and Data Relay satellite system.

Tera - 10^{12} or 2^{40}

Terabyte – 1012 bytes.

Test-and-set – coordination mechanism for multiple processes that allows reading to a location and writing it in a non-interruptible manner.

Thread – smallest independent set of instructions managed by a multiprocessing operating system.

T&I – test and integration.

TID – total ionizing dose.

TMR – triple modular redundancy.

Toolchain – set of software tools for development.

Transceiver – receiver and transmitter in one box.

Transducer – a device that converts one form of energy to another.

Train – a series of satellites in the same or similar orbits, providing sequential observations.

TRAP – exception or fault handling mechanism in a computer; an operating system component.

Triplicate – using three copies (of hardware, software, messaging, power supplies, etc.). for redundancy and error control.

TRL – technology readiness level

Truncate – discard. cutoff, make shorter.

TT&C – tracking, telemetry, and command.

ttl – transistor-transistor logic integrated circuit.

UART – Universal asynchronous receiver-transmitter.

UDP – User datagram protocol; part of the Internet Protocol.

uM – micro (10^{-6}) meter

Underflow – the result of an arithmetic operation is smaller than the smallest representable number.

UoSat – a family of small spacecraft from Surrey Space Technology Ltd. (UK).

USAF – United States Air Force.

USB – universal serial bus.

VDC – volts, direct current.

Vector – single dimensional array of values.

VHDL – very high level design language.

VIA – vertical conducting pathway through an insulating layer.

Virtual memory – memory management technique using address translation.

Virtualization – creating a virtual resource from available physical resources.

Virus – malignant computer program.

VLIW – very long instruction word – mechanism for parallelism.

VxWorks – real time operating system from Wind River systems.

WiFi – short range digital radio.

Watchdog – hardware/software function to sanity check the hardware, software, and process; applies corrective action if a fault is detected; fail-safe mechanism.

Wiki – the Hawaiian word for "quick." Refers to a collaborative content website.

Word – a collection of bits of any size; does not have to be a power of two.

Write-back – cache organization where the data is not written to main memory until the cache location is needed for re-use.
Write-through – all cache writes also go to main memory.
X-band – 7 – 11 GHz.
Xilinx – manufacturer of programmable logic and FPGA's.
Zener – voltage reference diode.
Zero address – architecture using implicit addressing, like a stack.
Zombie-sat – a dead satellite, in orbit.
Zone of Exclusion – volume in which the presence of an object, personnel, or activities are prohibited.

References

Antunes, Dr. Sandy, *Surviving Orbit the DIY Way: Testing the Limits Your Satellite Can and Must Match,* 2012 , Maker Media, Inc., ISBN-449310621.

Antunes, Dr. Sandy, *DIY Instruments for Amateur Space: Inventing Utility for Your Spacecraft Once It Achieves Orbit,* Maker Media, Inc., ISBN-978-1449310646.

Antunes, Dr. Sandy, *DIY Comms and Control for Amateur Space: Talking and Listening to Your Satellite,* 2015, Maker Media, Inc., ISBN-978-1449310660.

Antunes, Dr. Sandy, *DIY Satellite Platforms: Building a Space-Ready General Base Picosatellite for Any Mission,* 2012, Maker Media, Inc. ISBN-978-1449310608.

Ardila, David R. "Cubesats for Astrophysics," Aerospace Corp. https://cor.gsfc.nasa.gov/copag/rfi/CubesatsforAstrophysics.pdf

Asmar, Sami; Matousek, Steve "Mars Cube One (MarCO), First Planetary CubeSat Mission (presentation), 2014, JPL, avail: www.jpl.nasa.gov/cubesat/missions/marco.php

Avery, Keith; Fenchel, Jeffery; Mee, Jesse; Kemp, William; Netzer. Richard; Elkins, Donald; Zufelt, Brian; Alexander, David; *"Total Dose Test Results for Cubesat Electronics,"* 2011 IEEE Radiation Effects Data Workshop, 25-29 July 2011, Las Vegas, NV, pp. 1-8, 978-1-4577-1281-4.   avail: availwww.cosmiacpubs.org/pubs/TDTRCE.pdf

Barron, Andrew *Amsats and Hamsats: Amateur Radio and other Small Satellites,* 2018, ASIN- B07CG8DYR2.

Baisamo, James M. et al "CubeSat technology adaption for in-situ characterization of NEOs," presentation, avail: NASA Technical Reports Server (NTRS), 2014, document id 20140004799.

Betancourt, Mark "CubeSats to the Moon (Mars and Saturn, too)", Air & Space Magazine, Sept 2014.

Cudmore, Alan NASA/GSFC's Flight Software Architecture: core Flight Executive and Core Flight System, NASA/GSFC Code 582.

Cudmore, Alan Pi-Sat: A Low Cost Small Satellite and Distributed Mission Test Program, NASA/GSFC Code 582, https://ntrs.nasa.gov/archive/nasa/casi.ntrs.nasa.gov/20150023353.pdf

Datta, Lakshya Vaibhav, Guven Ugur (Ed) *Introduction to Nanosatellite Technology and Components: Applications of Cubesat Technology,* 2012, Lambert Academic Publishing, ISBN-978-3847314196.

DeCoursey, R.; Melton, Ryan; Estes, Robert R. Jr. "Sensors, Systems, and Next-Generation Satellites X," Proceedings of the SPIE, Vol. 6361 pp 63611m (2006).   (use of non-radiation hardened cpus).

De Selding Peter B. "1 in 5 Cubesats Violates International Disposal Guidelines," 2015. available, http://spacenews.com/1-in-5-Cubesats-violate-international-orbit-disposal-guidelines/

Dreißas, Riccardo *Cubesat electrical power system simulation: a generic approach*, 2013, AV Akademikerverlag, ISBN-978-3639473766.

Dunn, William R. *Practical Design of Safety-Critical Computer Systems*, July 2002, Springer, ISBN-0971752702.

Eickhoff, Jens *Onboard Computers, Onboard Software and*

Satellite Operations: An Introduction, 2011, Springer Aerospace Technology, ISBN-3642251692.

Engineering and Medicine, National Academies of Sciences (Author), Division on Engineering and Physical Sciences (Author), Space Studies Board (Author), *Achieving Science with Cubesats: Thinking Inside the Box,* National Academies Press, 2016, ISBN-978-0309442633.

Fortescue, Peter and Stark, John *Spacecraft System Engineering*, 2nd ed, Wiley, 1995, ISBN 0-471-95220-6.

Fowler, Kim *Mission-Critical and Safety-Critical Systems Handbook: Design and Development for Embedded Applications,* Newnes; 1st edition, November 20, 2009, ISBN-0750685670.

Gilster, Paul "CubeSats: Deep Space Possibilities," Sept. 2015, avail: http://www.centauri-dreams.org/?p=34056.

Griffith, Robert C. *Mobile CubeSat Command and Control (MC3)*, Feb. 2012, Amazon Digital Services, ASIN B007B4LWBBO.

Harland, David M. and Lorenz, Ralph D. *Space Systems Failures, Disasters and Rescues of Satellites, Rockets and Space Probes,*. 2005, Praxis Publishing, ISBN 0-387-21519-0.

Holmes-Siedle, A. G. and Adams, L. *Handbook of Radiation Effects*, 2002, Oxford University Press, ISBN 0-19-850733-X.

Kalinsky, David, "Architecture of Safety-Critical Systems," Aug 23. 2005, available white paper at www.embedded.com.

Kerber, Jonathas G. "An Introduction to Cubesats as Teaching Tools and Technology Testing Platforms, University of Ottawa.

Klofas, Bryan, "Frequency Allocation for Government-funded

Cubesats: NSF Paves the Way, 2011, avail: klofas.com/papers/klofas_nsf_comm_amsat2011.pdf

Krämer, Bernd J. and Völker, Norbert (Eds.) *Safety-Critical Real-Time Systems*, 2010, ISBN-10-1441950192.

Kreck Institute of Space Studies, *Small Satellites: A revolution in Space Science*, Final Report, July 2014. avail: kiss.caltech.edu/study/smallsat/KISS-SmallSat-FinalReport.pdf

Lappas, Vaios, et al *CubeSail: A low cost Cubesat based solar sail demonstration mission*, Advances in Space Research, 2001, 48.11, 1890-1901.

Logsdon, Tom *Orbital Mechanics: Theory and Applications*, 1997, Wiley-Interscience, ISBN 0471146366.

Mahdi, Mohammed Chessab *Introduction to Cubesat Technology and Subsystem: Orbit Design, Debris Impact, and Orbital Decay Prediction,* Lambert Academic Publishing, 2016, ISBN-978-3659960710.

Madni, Mohamed Atef; Raad, Raad; Tubbal, Faisal "Inter-CubeSat Communications: Routing Between CubeSat Swarms in a DTN Architecture," presentation, avail:https://icubesat.org/papers/2015-2/2015-b-2-1.

Maurer, Richard H.; Fraeman, Martin E.; Martin, Mark N.; Roth, David R. "Harsh Environments: Space Radiation Environment, Effects, and Mitigation, The Johns Hopkins University, Applied Physics Laboratory, Technical Digest, Vol 28, No. 1, 2008.

Messenger, G. C. and Ash, M. S. *The Effects of Radiation on Electronic Systems,* 1992, Van Nostrand Reinhold.

Mikaelian, Tsoline "Spacecraft Charging and Hazards to

Electronics in Space" May 2001, avail:
https://www.researchgate.net/publication/45858016_Spacecraft_C
harging_and_Hazards_to_Electronics_in_Space.

NASA, "Hitchhiking Into the Solar System: Launching NASA's
First Deep-Space Cubesats," avail: www.nasa.gov/exploration.

Petersen, Edward *Single Event Effects in Aerospace* 1st Ed. Wiley-
IEEE Press; 1 edition, October 4, 2011, ISBN- 0470767499.

Petro, Andrew "Small Spacecraft Technology, Briefing to NASA
Advisory   Committee,   April   15,   2014,"
https://www.nasa.gov/sites/default/files/files/APetro_SmallSpacecr
aft.pdf .

Prussing, John R and Conway, Bruce A. *Orbital Mechanics*, 1993,
Oxford University Press, ISBN- 0195078349.

Sahu, Kusum, *EEE-INST-002, Instructions for EEE Parts
Selection, Screening, Qualification, and Derating*, with addendum
1, April 2008, NASA/TP-2003-212242.

Schaire, Scott "NASA Near Earth Network (NEN) and Space
Network (SN) Support of Cubesat Communications, March, 2016,
avail: https://ntrs.nasa.gov/search.jsp?R=20160005439

Simmons, Kevin L; Christenson, Shawna L. "Let's Go To Space,
Building a Cubesat Team through Blue-Sky Learning, ISBN-
9798570577701.

Staehle, Robert et al, "Interplanetary Cubesats: Opening the Solar
System to a Broad Community at Lower Cost" Cubesat Workshop,
2011, Logan Utah. Avail:
https://www.nasa.gov/pdf/716078main_Staehle_2011_PhI_Cubesa
t.pdf

Stakem, Patrick H.   "Linux in Space", Oct. 2, 2003, invited presentation, Institution of Electrical Engineers, Sheffield Hallam University, Sheffield, UK.

Stakem, Patrick H. "The Applications of Computers and Microprocessors Onboard Spacecraft, NASA/GSFC 1980.

Stakem, Patrick H. "Free Software in Space–the NASA Case," invited paper, Software Livre 2002, May 3, 2002, Porto Allegre, Brazil.

Stakem, Patrick H., Korol, Guilherme, Gomes, Gabriel Augusto, "A Lightweight Open Source Command and Control Center and its Interface to Cubesat Flight Software." Presented at Flight Software-15 Conference (FSW-15), The Johns Hopkins University, Applied Physics Laboratory.

Stakem, Patrick H.; Rezende, Aryadne; Ravazzi, Andre "Cubesat Swarm Communications," 2016.

Stakem, Patrick H.; Da Costa, Rodrigo Santos Valente; Rezende, Aryasdne; Ravazzi, Andre "A Cubesat-based alternative for the Juno Mission to Jupiter, 2017, available from the author, pstakem@jhu.edu.

Stakem, Patrick H., Kerber, Jonathas "Rad-hard software, Cubesat Flight Computer Self-monitoring, Testing, Diagnosis, and Remediation," 2017, available from the author, pstakem1@jhu.edu.

Stakem, Patrick H. "Lunar and Planetary Cubesat Missions," March Volume 15, Polytech Revista de Tecnologia e Ciência, avail: http://www.polyteck.com.br/revista_online/ed_15.pdf

Swartwout, Michael *The First One Hundred Cubesats: A Statistical Look*, JoSS, Vol. 2, No. 2, pp. 213-233.

Swartwout, Michael "The Long-Threatened Flood of University-

Class Spacecraft (and Cubesats) Has Come: Analyzing the Numbers"
Avail: digitalcommons.usu.edu/cgi/viewcontent.cgi?article=2972&context=smallsat

Truszkowski, Walt; Hallock, Harold L.; Rouff, Christopher; Karlin, Jay; Rash, Hinchey, Mike; Sterritt, Roy *Autonomous and Automatic Systems,: With Applications to NASA Intelligent Spacecraft Operations and Exploration* Systems, Monographs in System and Software Engineering, Springer, 009, ISBN 9781846282331.

Truszkowski, Walt; Clark, P. E.; Curtis, S.; Rilee, M. Marr, G. A*NTS: Exploring the Solar System with an Autonomous Nanotechnology Swarm.* J. Lunar and Planetary Science XXXIII (2002).

Violette, Daniel P. "Arduino/Raspberry Pi: Hobbyist Hardware and Radiation Total Dose Degradation," 2014, presented at the EEE Parts for Small Missions Conference, NASA-GSFC, Greenbelt, MD, September 10-11, 2014.

Welti, C. Robert, , *Satellite Basics for Everyone: An Illustrated Guide to Satellites for Non-Technical and Technical People,* iUniverse, 2012, ASIN-B008I21XW6.

Wertz, James R. (ed) *Spacecraft Attitude Determination and Control*, section 6.9, Onboard Computers, 1980, Kluwer, ISBN 90-277-1204-2.

Wichmann, Brian A. *Software in Safety Related Systems*, Wiley, 1992. ISBN 0471-93474-7.

Wooster, Paul; Boswell, David; Stakem, Patrick; Cowan-Sharp, Jessy "Open Source Software for Small Satellites," SSC07-XII-3, 21st. Annual AIAA/USU, paper SSC07-XII-3, July 2007.

Resources

Small Spacecraft Technology State of the Art, NASA-Ames, NASA/TP2014-216648/REV1, July 2014.

Core Flight System (CFS) Deployment Guide, Ver. 2.8, 9/30/2010, NASA/GSFC 582-2008-012.

NASA Systems Engineering Handbook, NASA/SP-2007-6105, Rev. 1. (available on Google Books, Amazon, and others)

Cubesat Design Specification, Cubesat Program, California Polytechnic State University, avail:
https://www.google.com/search?
q=Cubesat+Design+Specification&ie=utf-8&oe=utf-8
and at www.Cubesat.org

Cubesat Concept and the Provision of Deployer Services, avail:
https://eoportal.org/web/eoportal/satellite-
missions/content/-/article/Cubesat-concept-1

www.ccsds.org

https://nasasearch.nasa.gov/

Spacetrack.org (requires an account)

http://www.celestrak.com/NORAD/elements/

http://satellitedebris.net/Database/

NASA, Software Documentation Standard, NASA-STD-2100-91, available:
https://ntrs.nasa.gov/archive/nasa/casi.ntrs.nasa.gov/19980228459.
pdf

Interplanetary Cubesats: Opening the Solar System to a Broad Community at Lower Cost; JPL, 2012, avail: https://www.nasa.gov/pdf/716078main_Staehle_2011_PhI_**Cubes at**.pdf

Cubesat: A new Generation of Picosatellite for Education and Industry Low-Cost Space Experimentation, 14th Annual/USU Conference on Small Satellites.
Avail: digitalcommons.usu.edu/cgi/viewcontent.cgi?article=2069&context=smallsat

MakerSat: A Cubesat Designed for In-Space 3D Print and Assembly, SSC16-WK-29,
avail: digitalcommons.usu.edu/cgi/viewcontent.cgi?article=3444&context=smallsat

Cubesat Design Specification (CDS) Rev. 13, Cubesat Program, Cal Poly. Avail: http://www.Cubesat.org/

NASA, John F. Kennedy Space Center, Launch Services Program, Program Level Dispenser and Cubesat Requirements Document, LSP-REQ-317.01, Rev. B, Jan. 2014.

Guidance on Obtaining Licenses for Small Satellites, Federal Communications Commission, March 15, 2013, 13-445. avail: https://www.fcc.gov/document/guidance-obtaining-licenses-small-satellites

Amateur Radio Satellite Organization (AMSAT) – www.amsat.org

Launch Services Program, Program Level Dispenser and Cubesat Requirements Document, NASA, John F. Kennedy Space Center, LSP-REQ-317.01, Rev. B. avail: www.nasa.gov/pdf/627972main_LSP-REQ-317_01A.pdf

General Payload Users Guide, Spaceflight, Inc. SF-2100-PUG-00001, www.spaceflightindustries.com

NASA Open Source Agreement, avail:
https://opensource.org/licenses/NASA-1.3

InterPlanetary Networking Special Interest Group (IPNSIG) -
http://ipnsig.org/

CubeSat: A new Generation of Picosatellite for Education and
Industry Low-Cost Space Experimentation, avail:
users.csc.calpoly.edu/~csturner/ssc01.pdf

Rapid Build and Space Qualification of Cubesats, avail:
www.digitalcommons.usu.edu/cgi/viewcontent.cgi?
article=1148&context=smallsat

Open Source Engineering Tools
 http://wiki.developspace.net/Open_Source_Engineering_Tools

100 Earth Shattering Remote Sensing Applications and Uses, 2015,
GIS Geography avail: http://gisgeography.com/100-earth-remote-
sensing-applications-uses/

Report Concerning Space Data Systems Standards, Mission
Operations Services Concept, CCSDS Informational Report,
CCSDS 520.0-G-3, Green Book, December 2010, avail, ccsds.org

Overview of Space Communications Protocols, avail:
 cwe.ccsds.org/sls/docs/SLS.../130x0g2_master_Dec16_2013.docx
Core Flight System – http://cfs.gsfc.nasa.gov`

http://pmddtc.state.gov/

http://srag-nt.jsc.nasa.gov/SpaceRadiation/What/What.cfm

https://www.nasa.gov/content/goddard/the-future-of-cubesats,
2017.

https://www.esa.int/Enabling_Support/Preparing_for_the_Future/Discovery_and_Preparation/CubeSats

www.cubesat.org

https://www.nasa.gov/content/goddard/the-future-of-cubesats

https://www.esa.int/Enabling_Support/Preparing_for_the_Future/Discovery_and_Preparation/CubeSats

NASA, Cubesat101, Basic Concepts and Processes for First-Time CubeSat Develoers. Oct. 2017. avail:

https://www.nasa.gov/sites/default/files/atoms/files/nasa_csli_cubesat_101_508.pdf

https://sites.google.com/a/slu.edu/swartwout/home/cubesat-database

https://www.nasa.gov/feature/first-cubesat-built-by-an-elementary-school-deployed-into-space

https://www.researchgate.net/publication/311279467_BOOM_OF_THE_CUBESAT_A_STATISTIC_SURVEY_OF_CUBSATS_LAUNCH_IN_2003-2015/download

wikipedia, various

Country's that have launched Cubesats

as of 9/18/21

This list was complied in September of 2021, and may not be complete. This lists the countrys that launched Cubesats as a part of their National Space Program, or by a University in the Country.

1. Australia
2. Bangladesh
3. Belgium
4. Brazil
5. Canada
6. China
7. Costa Rica
8. Denmark
9. Ecuador
10. Ethiopia
11. Finland
12. France
13. Galacia (autonomous area of Spain)
14. Germany
15. Ghana
16. Greece
17. India
18. Italy
19. Japan
20. Jourdan
21. Kenya

22. Korea (south)
23. Lithuania
24. Malaysia
25. Mongolia
26. Netherlands
27. New Zealand
28. Nigeria
29. Norway
30. Pakistan
31. Peru
32. Poland
33. Puerto Rico
34. Romania
35. Singapore
36. Slovakia
37. South Africa
38. Spain
39. Switerland
40. Tunsia
41. Turkey
42. UK
43. Uruguay
44. USA
45. Vietnam

If you enjoyed this book, you might also be interested in some of these.

Stakem, Patrick H. *16-bit Microprocessors, History and Architecture*, 2013 PRRB Publishing, ISBN-1520210922.

Stakem, Patrick H. *4- and 8-bit Microprocessors, Architecture and History*, 2013, PRRB Publishing, ISBN-152021572X,

Stakem, Patrick H. *Apollo's Computers,* 2014, PRRB Publishing, ISBN-1520215800.

Stakem, Patrick H. *The Architecture and Applications of the ARM Microprocessors,* 2013, PRRB Publishing, ISBN-1520215843.

Stakem, Patrick H. *Earth Rovers: for Exploration and Environmental Monitoring,* 2014, PRRB Publishing, ISBN-152021586X.

Stakem, Patrick H. *Embedded Computer Systems, Volume 1, Introduction and Architecture*, 2013, PRRB Publishing, ISBN-1520215959.

Stakem, Patrick H. *The History of Spacecraft Computers from the V-2 to the Space Station*, 2013, PRRB Publishing, ISBN-1520216181.

Stakem, Patrick H. *Floating Point Computation*, 2013, PRRB Publishing, ISBN-152021619X.

Stakem, Patrick H. *Architecture of Massively Parallel Microprocessor Systems*, 2011, PRRB Publishing, ISBN-1520250061.

Stakem, Patrick H. *Multicore Computer Architecture,* 2014, PRRB

Publishing, ISBN-1520241372.

Stakem, Patrick H. *Personal Robots*, 2014, PRRB Publishing, ISBN-1520216254.

Stakem, Patrick H. *RISC Microprocessors, History and Overview,* 2013, PRRB Publishing, ISBN-1520216289.

Stakem, Patrick H. *Robots and Telerobots in Space Application*s, 2011, PRRB Publishing, ISBN-1520210361.

Stakem, Patrick H. *The Saturn Rocket and the Pegasus Missions, 1965,* 2013, PRRB Publishing, ISBN-1520209916.

Stakem, Patrick H. *Visiting the NASA Centers, and Locations of Historic Rockets & Spacecraft,* 2017, PRRB Publishing, ISBN-1549651205.

Stakem, Patrick H. *Microprocessors in Space*, 2011, PRRB Publishing, ISBN-1520216343.

Stakem, Patrick H. Computer *Virtualization and the Cloud*, 2013, PRRB Publishing, ISBN-152021636X.

Stakem, Patrick H. *What's the Worst That Could Happen? Bad Assumptions, Ignorance, Failures and Screw-ups in Engineering Projects, 2014,* PRRB Publishing, ISBN-1520207166.

Stakem, Patrick H. *What's the Worst That Could Happen? More Bad Assumptions, Ignorance, Failures, and Screw-ups in Engineering Projects, Vol. 2,* ISBN-978-1981005574

Stakem, Patrick H. *Computer Architecture & Programming of the Intel x86 Family, 2013,* PRRB Publishing, ISBN-1520263724.

Stakem, Patrick H. *The Hardware and Software Architecture of the Transputer,* 2011,PRRB Publishing, ISBN-152020681X.

Stakem, Patrick H. *Mainframes, Computing on Big Iron*, 2015, PRRB Publishing, ISBN- 1520216459.

Stakem, Patrick H. *Spacecraft Control Centers*, 2015, PRRB Publishing, ISBN-1520200617.

Stakem, Patrick H. *Embedded in Space,* 2015, PRRB Publishing, ISBN-1520215916.

Stakem, Patrick H. *A Practitioner's Guide to RISC Microprocessor Architecture*, Wiley-Interscience, 1996, ISBN-0471130184.

Stakem, Patrick H. *Cubesat Engineering*, PRRB Publishing, 2017, ISBN-1520754019.

Stakem, Patrick H. *Cubesat Operations*, PRRB Publishing, 2017, ISBN-152076717X.

Stakem, Patrick H. *Interplanetary Cubesats*, PRRB Publishing, 2017, ISBN-1520766173 .

Stakem, Patrick H. Cubesat Constellations, Clusters, and Swarms, Stakem, PRRB Publishing, 2017, ISBN-1520767544.

Stakem, Patrick H. *Graphics Processing Units, an overview*, 2017, PRRB Publishing, ISBN-1520879695.

Stakem, Patrick H. *Intel Embedded and the Arduino-101, 2017,* PRRB Publishing, ISBN-1520879296.

Stakem, Patrick H. *Orbital Debris, the problem and the mitigation*, 2018, PRRB Publishing, ISBN-*1980466483.*

Stakem, Patrick H. *Manufacturing in Space*, 2018, PRRB Publishing, ISBN-1977076041.

Stakem, Patrick H. *NASA's Ships and Planes*, 2018, PRRB Publishing, ISBN-1977076823.

Stakem, Patrick H. *Space Tourism*, 2018, PRRB Publishing, ISBN-1977073506.

Stakem, Patrick H. *STEM – Data Storage and Communications*, 2018, PRRB Publishing, ISBN-1977073115.

Stakem, Patrick H. *In-Space Robotic Repair and Servicing*, 2018, PRRB Publishing, ISBN-1980478236.

Stakem, Patrick H. *Introducing Weather in the pre-K to 12 Curricula, A Resource Guide for Educators*, 2017, PRRB Publishing, ISBN-1980638241.

Stakem, Patrick H. *Introducing Astronomy in the pre-K to 12 Curricula, A Resource Guide for Educators*, 2017, PRRB Publishing, ISBN-198104065X.
Also available in a Brazilian Portuguese edition, ISBN-1983106127.

Stakem, Patrick H. *Deep Space Gateways, the Moon and Beyond*, 2017, PRRB Publishing, ISBN-1973465701.

Stakem, Patrick H. *Exploration of the Gas Giants, Space Missions to Jupiter, Saturn, Uranus, and Neptune*, PRRB Publishing, 2018, ISBN-9781717814500.

Stakem, Patrick H. *Crewed Spacecraft*, 2017, PRRB Publishing, ISBN-1549992406.

Stakem, Patrick H. *Rocketplanes to Space*, 2017, PRRB Publishing, ISBN-1549992589.

Stakem, Patrick H. *Crewed Space Stations,* 2017, PRRB Publishing, ISBN-1549992228.

Stakem, Patrick H. *Enviro-bots for STEM: Using Robotics in the pre-K to 12 Curricula, A Resource Guide for Educators,* 2017, PRRB Publishing, ISBN-1549656619.

Stakem, Patrick H. *STEM-Sat, Using Cubesats in the pre-K to 12 Curricula, A Resource Guide for Educators*, 2017, ISBN-1549656376.

Stakem, Patrick H. *Lunar Orbital Platform-Gateway*, 2018, PRRB Publishing, ISBN-1980498628.

Stakem, Patrick H. *Embedded GPU's*, 2018, PRRB Publishing, ISBN- 1980476497.

Stakem, Patrick H. *Mobile Cloud Robotics*, 2018, PRRB Publishing, ISBN- 1980488088.

Stakem, Patrick H. *Extreme Environment Embedded Systems,* 2017, PRRB Publishing, ISBN-1520215967.

Stakem, Patrick H. *What's the Worst, Volume-2*, 2018, ISBN-1981005579.

Stakem, Patrick H., *Spaceports*, 2018, ISBN-1981022287.

Stakem, Patrick H., *Space Launch Vehicles*, 2018, ISBN-1983071773.

Stakem, Patrick H. *Mars*, 2018, ISBN-1983116902.

Stakem, Patrick H. *X-86, 40th Anniversary ed*, 2018, ISBN-1983189405.

Stakem, Patrick H. *Lunar Orbital Platform-Gateway*, 2018, PRRB Publishing, ISBN-1980498628.

Stakem, Patrick H. *Space Weather*, 2018, ISBN-1723904023.

Stakem, Patrick H. *STEM-Engineering Process*, 2017, ISBN-1983196517.

Stakem, Patrick H. *Space Telescopes,* 2018, PRRB Publishing, ISBN-1728728568.

Stakem, Patrick H. *Exoplanets*, 2018, PRRB Publishing, ISBN-9781731385055.

Stakem, Patrick H. *Planetary Defense*, 2018, PRRB Publishing, ISBN-9781731001207.

Patrick H. Stakem *Exploration of the Asteroid Belt*, 2018, PRRB Publishing, ISBN-1731049846.

Patrick H. Stakem *Terraforming*, 2018, PRRB Publishing, ISBN-1790308100.

Patrick H. Stakem, *Martian Railroad, A Design Exercise,* 2019, PRRB Publishing, ISBN-1794488243.

Patrick H. Stakem, *Exoplanets,* 2019, PRRB Publishing, ISBN-1731385056.

Patrick H. Stakem, *Exploiting the Moon,* 2019, PRRB Publishing, ISBN-1091057850.

Patrick H. Stakem, *RISC-V, an Open Source Solution for Space Flight Computers,* 2019, PRRB Publishing, ISBN-1796434388.

Patrick H. Stakem, *Arm in Space*, 2019, PRRB Publishing, ISBN-9781099789137.

Patrick H. Stakem, *Extraterrestrial Life*, 2019, PRRB Publishing, ISBN-978-1072072188.

Stakem, Patrick H. Submarine Launched Ballistic Missiles, 2019, ISBN-978-1088954904.

Patrick H. Stakem, *Space Command*, Military in Space, 2019, PRRB Publishing, ISBN-978-1693005398.

Stakem, Patrick H. *Enviro-bots for STEM: Using Robotics in the pre-K to 12 STEM Curricula, A Resource Guide for Educators,* 2017, ISBN- 978-1549656613.

Patrick H. Stakem, *Exploration of Lunar & Martian Lava Tubes by Cube-X*, ISBN 979-8621435325.

Patrick H. Stakem, *Robotic Exploration of the Icy moons of the Gas Giants*, 2020, PRRB Publishing, ISBN- 979-8621431006.

Stakem, Patrick H. *Hacking Cubesats*, 2020, PBRRB Publishing, ISBN-979-8623458964.

Powerships, Powerbarges, Floating Wind Farms: electricity when and where you need it, 2021, PRRB Publishing, ISBN-979-8716199477.

Hospital Ships, Trains, and Aircraft, 2020, PRRB Publishing, ISBN-979-8642944349.

2020/2021 Releases

CubeRovers, a Synergy of Technologys, 2020, ISBN-979-8651773138

Exploration of Lunar & Martian Lava Tubes by Cube-X, ISBN-979-8621435325.

Robotic Exploration of the Icy moons of the Gas Giants, ISBN-979-8621431006.

History & Future of Cubesats, ISBN-978-1986536356.

Robotic Exploration of the Icy Moons of the Ice Giants, by Swarms of Cubesats, ISBN-979-8621431006.

Swarm Robotics, ISBN-979-8534505948.

Introduction to Electric Power Systems, ISBN-979-8519208727.

Centros de Control: Operaciones en Satélites del Estándar CubeSat (Spanish Edition), 2021, ISBN-979-8510113068.

www.ingramcontent.com/pod-product-compliance
Lightning Source LLC
Chambersburg PA
CBHW020613220526
45463CB00006B/2578